NATIO
GEOGR

School Pu

FREAKY FROGS

PATHFINDER EDITION

By Dan and Michele Hogan

CONTENTS

FREAKY FROGS

Worldwide something weird is happening to frogs.

By Dan and Michele Hogan

Skinny or fat, smelly or poisonous, there are many kinds of frogs. Scientists are constantly finding new species, or kinds. And they're finding the amphibians all over the world.

Here are a few examples. Recently, tiny frogs, the size of a dime, were discovered under old leaves in Cuba. Parachuting frogs were spotted on Sumatra and Java. See-through "glass" frogs were found in Guyana. In Venezuela a frog that lets off a horrible smell when threatened was spotted.

You might think finding new frog species is good news. But not all frog news is rosy. Many frogs are hopping into very big problems.

If you look around some ponds and creeks, you might find that something is missing—frogs. Scientists have noticed the same thing. In many parts of the world, fewer frogs are hopping around than there used to be. Frogs are even vanishing from our country.

One lost species is the gastric brooding f rog in Australia. It had a special way of protecting its young. The mother frog kept her tadpoles in her stomach until they turned into baby frogs. Scientists now say the frog is extinct.

Many of the remaining frogs look odd. Some of them don't live very long.

Why are frogs dying out? No one knows for sure. Scientists say there may be many reasons. Let's look at some of them.

Blue poison frogs

Wet and Wild

Frogs are found on every continent except Antarctica. Some live in cities, deserts, mountains, or grasslands. But most frogs prefer wet areas. The wetter, the better. They really like ponds, marshes, and rain forests.

Frogs like wet areas for many reasons. They lay their eggs in water. They find their favorite meals—flies, snails, worms, and other tasty treats—there as well. They also find protection in water. A frog can hide from birds and other predators by darting under a nearby leaf or swimming underwater.

Some frogs are having a hard time finding soggy homes. Frog habitats are drying up. There are fewer wetlands today. Worldwide, people are draining ponds and marshes. They're building houses and businesses in those areas. As the new structures go up, frogs lose their old soggy homes.

Habitat loss isn't the only problem frogs face. Some experts say more animals are preying on the amphibians than ever before. In Australia, for example, people have stocked ponds with fish. Some of those fish like to eat frogs' eggs. As predators increase, frogs decrease.

Chemical Concerns

Another problem that worries scientists is the use of some chemicals. These can harm frogs.

Rain can wash the chemicals into ponds. The chemicals not only pollute the water but also seep into frogs' skin. After the chemicals enter a frog's body, they can cause big problems.

Some of the chemicals make tadpoles and young frogs eat less and swim more slowly. Other chemicals can change the way male frogs look and act. Some of the affected frogs produce fewer tadpoles, which means there are fewer frogs.

Daddy Day Care. A male frog in Papua New Guinea guards his young.

Looking Out.
Watchful eyes help a Peruvian tree frog spot both enemies and prey.

FAR-OUT FROGS

Smallest: The gold frog is about the size of your fingernail.

Biggest: The goliath frog is almost a foot long. It can weigh as much as a house cat.

Singer: A frog in China sings like a bird.

Survivor: The wood frog can stay frozen for several weeks. It goes back to normal life after thawing out.

Java flying frog

Many Mutants

Something even freakier is happening to frogs. Many are mutating, or changing. The first **mutant** frogs were discovered nearly 60 years ago. Then there were only a few mutant frogs. Now there are lots of them.

Deformed frogs have been found in 38 U.S. states and in many other countries. Nearly half the frogs in many areas are deformed in some way. The deformities aren't good for frogs.

Many of the frogs have no legs. Others have too many legs. Some have legs growing from strange places—even their stomachs. Most deformed frogs do not live as long as normal frogs. What's causing the changes?

5

Jump In. You may have seen this frog. The leopard frog is common in North America.

Bad Frog. This frog has too many legs. Something has caused it to mutate.

Rays From Space

So far scientists don't know what is causing the problems. They point to many possibilities. Pollution, diseases, and even **ultraviolet** (UV) light might cause the freaky features.

UV light can harm your skin. It can give you a sunburn and cause skin cancer. That's why it is important to wear sunblock whenever you go outside.

The dangerous rays are even more harmful to frogs. They can pass through frog eggs. That might harm the eggs. The damaged eggs could produce mutant frogs.

UV light is a bigger problem than it used to be. Normally a layer of Earth's air blocks most UV light. It's called the **ozone layer**. But a kind of pollution has damaged the layer. Now more UV light reaches Earth's surface.

Luckily the chemicals that damaged the ozone layer are no longer made. Scientists say the layer should slowly fix itself. But that will take many years. Once the hole is gone, UV light will not be as big a problem as it is today.

Deadly Disease

A disease might also cause the mutations. The disease is carried and spread by a tiny worm.

The worm is a parasite, a plant or animal that lives off of other creatures. If a worm enters a frog's body, it can give the disease to the frog. The disease causes extra or missing limbs in the frog's young.

The disease-carrying worm seems to be spreading. As the worm moves into new lakes and ponds, it brings the disease with it. That means more frogs can catch it.

Frogs are not the only creatures that can catch the disease. Some other amphibians, including toads and salamanders, can get it as well.

The Future of Frogs

Dying and mutating frogs may seem like a small problem. Think again. Scientists say that something much bigger is going on. Whatever is affecting frogs might one day affect other animals—and people—as well.

Disappearing frogs could also mean we will have fewer medicines in the future. Many frogs have poisons, or **toxins**. Scientists use those toxins to make medicines. One frog toxin is being used to make painkillers.

But there is hope for frogs. Scientists have cleaned ponds where frogs used to live. And the frogs returned!

Scientists aren't the only folks who can help frogs. You can too. What can you do?

Pull weeds out by hand instead of using spray. Avoid using fertilizers near streams or ponds. Keep your pets away from frogs.

Your work might even help other animals and people stay healthy too.

Green tree frogs

WORDWISE

amphibian: animal that lives in water while young and later on land

extinct: completely gone

mutant: abnormal

ozone layer: gases in Earth's atmosphere

species: type of animal or plant

toxin: poison

ultraviolet: type of invisible ray from the sun

TROUBLE

Frogs are not the only animals having problems in the wild. Newts, salamanders, and other amphibians are in trouble too. Why? It could have something to do with how they grow. You see, frogs and other amphibians go through some incredible changes in their lives. Their changes may relate to their problems.

The Cycle of Life

All animals change as they grow. But amphibians change more than most. Look at the pictures below to see how frogs grow and change. It's almost like they have two lives—one in water and one on land. That's what the word amphibian means: adapted to life in water and on land.

LIFE OF A FROG

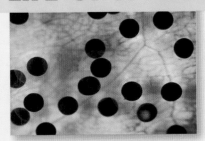

Eggs A frog begins life as one of many eggs in water. Inside the egg, the frog grows and changes. Soon it will break out of its shell and swim into the water.

Tadpole A young frog is called a tadpole. It has a tail and gills for breathing underwater. It lives in the water and swims in search of food.

Adult Over time, the tadpole grows legs. It loses its tail and develops lungs. Skin grows over its gills. The adult frog spends most of its time on land.

Double Trouble

Living in two places could mean double trouble for frogs and other amphibians. It means the critters can pick up pollution from the water, but also from the land. They can get it from the air they breathe too. Maybe this is one reason that frogs are turning freaky.

Sensitive Skin

Another reason amphibians are at risk may be their sensitive skin.

The skin of many animals is thick and strong. It keeps germs and other things out. But an amphibian's skin helps to bring things in.

For example, frogs don't drink. Instead, their skin lets in the moisture they need. Frogs can also breathe through their skin. This lets them take in extra oxygen as they swim.

Dangerous chemicals in the water can also pass through the animal's skin— and that can lead to some serious problems for the frog.

Amphibians With Answers

Whatever the reason, amphibians are among the first to feel the effects of pollution. They might also be part of the solution. Scientists can watch the creatures for any signs of change. At the first hint of trouble, people can work to clean an area up. It helps the frogs— and people too.

Looking Ahead. Keeping areas free of pollution can help amphibians, such as this red-legged frog, stay healthy.

FROGS

It's time to leap into the world of frogs and find out how much you learned.

1 Where do frogs usually live?

2 What can cause frogs to lose their habitats?

3 What problems are some frogs having?

4 How might the sun be affecting frogs?

5 Why else are frogs and other amphibians having trouble?